STEVE P

MW01709744

Discover & learn about

Australian

Forests

AND WOODLANDS

Text Pat Slater

Photographs Steve Parish

ARK AUSTRALIA Steve Parish

HABITATS & ECOSYSTEMS

Contents

see page 35

see page 20

see page 29

The voyage of Ark Australia

Once upon a time, there was only one piece of land in all the world. It was called Pangea, and about 160 million years ago it split into two supercontinents, Laurasia and Gondwana.

By 70 million years ago, India, New Zealand and South America had broken away from Gondwana, leaving Antarctica and Australia behind. By 40 million years ago, Australia was drifting northwards on its own.

Like a giant Noah's Ark, the new continent carried a cargo of living things. Over long periods of time, as climates changed, these life forms changed in order to survive. Marsupials[G] became the most important mammals[G]. From around 30 million years ago, eucalypts and other drought-resistant plants became the most common flora[G]. As Australia neared Asia, new animals and plants arrived by air or by sea.

Then, 60 000 or more years ago, a new crew boarded the Ark. The ancestors of today's Aboriginal Australians crossed the narrowing sea from Asia and rapidly spread over the continent. They changed the balance of animal and plant life by their hunting methods and use of fire.

Australia's Greater Glider is a marsupial that lives in the treetops and eats leaves. It fills the place that monkeys occupy in forests in Africa, South America and India.

Kangaroos and eucalypt trees are typical of Australia's animal and plant life.

In 1788, European humans arrived on the Ark, bringing the saw, the axe and the gun. With them came new animals, including cats and foxes, horses, sheep, cattle and pigs.

In just over 200 years, great changes took place. Forests and woodlands, rivers and wetlands were destroyed or altered forever.

Gradually, Australians have come to realise that the Ark's resources need to be saved.

We hope that some of the damage to its systems can be repaired and that some areas can be preserved as wilderness.

At present, Ark Australia still has on board much of its cargo of unique[G] living things. However, some of them are gone forever, and some survive only in tiny, protected areas. Some are still disappearing.

This series of books describes the places where plants and animals live in Australia today – the habitats of Ark Australia. This volume is on the continent's forests.

Forest ecology

ECOSYSTEMS, HABITATS AND NICHES

Ecology: The word ecology comes from Greek words meaning "the study of a household or home". Studying the ecology of a place is finding out how living things survive in their environment[G], taking things from it and giving things back to it. This book is about the ecology of Australia's forests.

Ecosystem: An ecosystem is a community[G] of organisms – living things such as plants and animals – interacting with each other and making use of non-living things such as water, air, sunlight and soil.

Habitat: Many plants and animals survive best in just one part of an ecosystem. This is their preferred habitat. In a forest there are many different habitats, such as ground litter, rotting logs, tree trunks and the leafy canopy[G].

Niche: Within a habitat there will be niches – special places that suit just a few particular species.

ONE FOREST, MANY HABITATS

The Emperor Gum Moth caterpillar lives in the canopy.

Climbing mammals include the Brush-tailed Phascogale.

Bettongs and other mammals feed in the ground litter.

THIS FOREST IS AN ECOSYSTEM. IN IT ARE MANY HABITATS. IN THEM ARE MANY ECOLOGICAL NICHES.

Plants such as Waratahs and eucalypts make food and energy from non-living things.

The Koala's ecological niche is in the leafy canopies of trees.

Many forest creatures use creeks for drinking and bathing.

Groups and names

SORTING LIVING THINGS

The earth is home to an immense number and variety of living things. Since the time of the Swedish naturalist Carl Linnaeus (1707–78), living things have been placed in groups of related things called taxa. These **taxa** are placed in ranks from **species** (lowest) to **kingdom** (highest).

Organisms of one **species** can breed with each other and produce fertile[G] offspring.

Related **species** are grouped into **genera** (singular genus).

Related **genera** are grouped into **families**.

Related **families** are grouped into **orders**.

Related **orders** are grouped into **classes**.

Related **classes** are grouped into **phyla** (animals, *singular* **phylum**) and **divisions** (plants).

Related **phyla** and related **divisions** are grouped into **kingdoms**.

THE FIVE KINGDOMS OF LIVING THINGS

Kingdom Monera: Bacteria, which have a simple, one-celled structure; e.g., cyanobacteria, rikettsias, chlamydias.

Kingdom Protista: Single-celled and many-celled organisms that are not fungi, plants or animals. This group ncludes many tiny organisms and algae. Seaweeds are algae.

Kingdom Fungi: Many-celled organisms that digest food outside, not inside, their bodies. They do not use the Sun's energy to make food. Fungi include mushrooms, moulds and lichens.

Kingdom Plantae: Many-celled, usually land-living organisms. Their cell walls consist of stiff material called cellulose. A plant's tissues contain green chlorophyll. Chlorophyll allows the plant to use sunlight to make food in a process called photosynthesis. Plants include ferns, conifers and flowering plants.

Kingdom Animalia: Many-celled organisms whose cell walls are far less rigid than plant cells'. Animals cannot make food by photosynthesis. In the end, animals depend on plants for nourishment. Animals include sponges, insects, worms, birds and mammals such as humans.

Kingdom Monera: cyanobacteria.

Kingdom Protista: algae (seaweed).

Kingdom Fungi: fungus (toadstool).

Kingdom Plantae: plant (Mountain Devil).

Kingdom Animalia: animal (Eastern Rosella).

CLASSIFICATION OF THE DINGO *Canis lupus dingo*

Kingdom: Animalia (animals)
Phylum: Chordata (animals with backbones)
Class: Mammalia (warm-blooded, hairy-skinned chordates that feed their babies on milk)
Order: Carnivora (mammals that eat flesh)
Family: Canidae (dog-like carnivores[G])
Genus: *Canis* (dogs)
species: *lupus dingo* (the *dingo* subspecies of the Indian Wolf, *Canis lupus*)

An organism's scientific name, given only to that organism, consists of its genus and species, in italic type. So the Dingo is *Canis lupus dingo*, and cannot be confused with any other creature.

A Dingo (*Canis lupus dingo*).

Biodiversity

ARK AUSTRALIA'S RICH CARGO

The Greek word for life is *bios*. The biodiversity of an area is the range of different living things found in that area.

The greater the number of habitats in an ecosystem, the greater the biodiversity.

Australia is rich in biodiversity, and forests are amongst the continent's richest habitats. They are home to many species found nowhere else.

The biggest threat to Australia's biodiversity is the clearing of habitats by humans and their domestic stock. The next biggest threat comes from feral[G] species of animals and plants. Other pressures include pollution[G], the increase in salt in many places and the warming of the world's climates.

REASONS FOR AUSTRALIA'S BIODIVERSITY

- For most of its history, Australia has been either at the end of a chain of linked continents or on its own. It has been like a huge, lonely laboratory in which life forms could evolve[G].
- When Australia broke away from Antarctica, it carried no monkeys, apes, cats, dogs, cattle, sheep or goats. Marsupials and reptiles[G] have filled many ecological niches that are filled by these placental[G] mammals in other lands.
- Australia does not have a history of violent earth movements. The weather systems that maintain habitats have changed slowly over long periods of time. These facts have given living things time to evolve or change their behaviour to meet new conditions.

TWO AUSTRALIAN AREAS THAT ARE VERY RICH IN SPECIES

The south-west of Western Australia is rich in plant species.

Rose Banjine.

Catspaw.

Splendid Spider Orchid.

Yellow Flag.

Swan River Myrtle.

Mottlecah.

Fringed Lily.

Golden Dryandra.

The tropical rainforests of Queensland are rich in tree-living marsupials.

Lumholtz's Tree-kangaroo.

Striped Possum.

Spotted Cuscus.

Common Brushtail Possum.

Long-tailed Pygmy-possum.

Green Ringtail Possum.

Lemuroid Ringtail Possum.

Herbert River Ringtail Possum.

FACTS 'N' FIGURES FILE

- **Eighty-five per cent** of Australia's native[G] plant species and 82% of native mammal species are endemic (found nowhere else).
- **Australia has more endemic families** of flowering plants than any other country.
- **Areas of Australia** that are particularly species-rich are the Queensland rainforests and south-west Western Australia.
- **Fossils[G] show that the ancestors** of some Australian pines and cycads lived 175 million years ago, when Pangea was the only landmass on earth.
- **Many Australian rainforest trees** have changed little since the continent split from Gondwana.
- **Green algae,** which grow in salt or fresh water or in damp soil, include the closest relatives of land plants.
- **In drought times,** many Australian mosses appear dead. As soon as rain falls, they soak up water and come to life again.
- **The fossil history of the Queensland lungfish** can be traced back 400 million years.

The kingdom of plants

All life on earth, except for some bacteria, depends on plants for food, and for the oxygen that they release. Plants contain chlorophyll, a green substance that lets them use the power of sunlight to make food and energy in a process called photosynthesis. The raw materials needed include carbon (taken from the air as carbon dioxide), water and nitrogen[G].

SOME TYPES OF PLANTS FOUND IN AUSTRALIAN FORESTS

• **Mosses** are small, soft plants that do not have special vessels[G] inside them to carry liquids. They soak up water in the same way that sponges do.

• **Ferns** have special vessels, or vascular systems, that carry liquids through their tissues. Their bodies are supported by stiff material called lignin. They reproduce[G] by spores[G].

• **Seed plants** have vascular systems and stiff supportive tissue. They reproduce by means of female cells (ovules) and male cells (pollen). These fuse and develop into seeds that grow into new plants. Seed plants found in Australia include cycads, conifers and flowering plants.

Cycads are the most ancient of living seed plants. Their fossil record extends back to early in the Permian (290–245 million years ago). Australia has 23 species. Cycad roots may store food. Special roots contain cyanobacteria that can make nitrogen gas available for the plant to use. The ovules and pollen of cycads are produced in female and male cones.

Conifers are the most widespread non-flowering seed plants. They were the world's dominant plants in the Mesozoic (245–65 million years ago). Conifer pollen and ovules are borne in cones. Hoop, Huon, Bunya, Callitris, Norfolk Island and Kauri Pines are conifers.

Flowering plants are the dominant plants of today. They come in various forms and sizes, from floating duckweed to the tall Mountain Ash, and are placed in two groups:

Monocotyledons have one seed-leaf waiting to sprout inside the seed (e.g. a grain of wheat). They are non-woody, the veins in their leaves run beside each other rather than in a network and their flower parts are in threes. They include grasses, lilies and palms.

Dicotyledons have two seed-leaves curled inside the seed (e.g. a pea). They are woody trees, bushes or herbs[G], their leaf veins form networks, and their flower parts are in fours or fives. They include magnolias, daisies, peas, banksias and eucalypts.

Moss on a tree trunk.

Tree ferns.

A cycad (male cones of Riedle's Macrozamia).

A conifer (Bunya Pine).

A monocotyledon (Kangaroo Paw).

A dicotyledon (Finke River Mallee).

Types of forest

Forests cover about 5% of Australia. They grow mainly around the edges of the continent, on areas that receive more than 800 mm of rain each year. On the inland, drier edges of the forests, the trees become spaced further apart and their crowns become rounded. The spaces between crowns become wider, and eventually the forest becomes a woodland.

A forest consists of trees and smaller plants. A forest ecosystem will contain many animals as well.

FOREST TYPES

Forests are divided into different types by measuring three things:

- The completeness with which the trees' crowns cover the ground below them.
- The height of the tallest and commonest trees.
- The number of plant layers within the forest.

Trees can be grouped into three height categories:

- **low: 2–10 m**
- **medium: 11–30 m**
- **tall: greater than 30 m**

CLOSED FORESTS

Rainforests and mangrove forests are made up of trees whose crowns shade 81–100% of the ground beneath. This closed canopy may be more than 40 m above the ground.

Australia's rainforests can be divided into tropical (northern) and temperate (southern) types. They may overlap, but rarely mix. This is because, in the tropics, temperate rainforest only occurs on cool mountain tops; in the south, tropical rainforest survives only in warm, damp gullies. Rainforests do not survive fierce fires.

The canopy of a closed forest blocks out a view of the ground below. This is an aerial view of tropical rainforest near Cairns, Q.

OPEN FORESTS

Eucalypt, acacia, melaleuca, callitris and casuarina forests shade 51–80% of the ground beneath. They are made up of sclerophyllous trees with stiff, waxy or leathery leaves.

Australia's sclerophyll forests can be divided into wet and dry forests. Wet sclerophyll forests are dominated by tall eucalypts that grow back from seed after fierce fires. Dry sclerophyll forests survive fires. Their trees sprout again from special roots or under-bark buds. Some dry sclerophyll forest plants have seeds that need fire to sprout.

The leafy canopy of an open forest does not block out a view of the sky above. This eucalypt forest is in the south-west of Western Australia.

WOODLANDS

When the tree crowns shade only 20–50% of the ground beneath, the area is called woodland. It may contain tall trees, but they are spaced far apart, like trees in a park.

"BUSH" and "SCRUB"

The term bush, or bushland, is often used to refer to open forests and woodlands.

Scrub, or jungle, was an old term for closed forests.

Where forests are

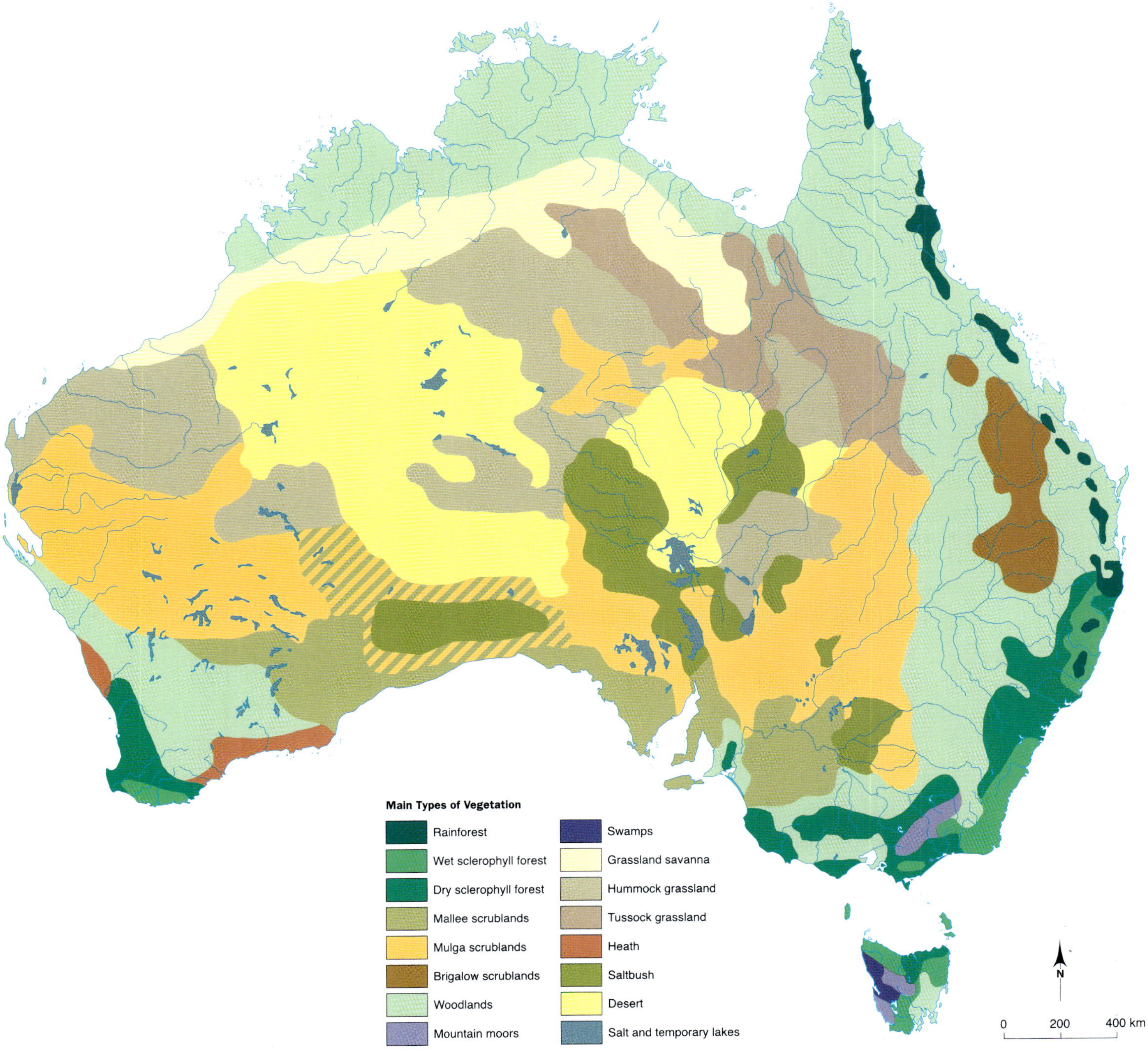

Rainforests

TROPICAL AND SUBTROPICAL

Above: Tropical rainforest in Queensland's Wet Tropics World Heritage Area.
Right, top–bottom: Male Eclectus Parrot; Green Python.

Above, top–bottom: Male Australian King-Parrot; Southern Forest Dragon.
Above right: Subtropical rainforest in Lamington National Park, south-east Queensland.

Rainforests

TROPICAL AND SUBTROPICAL

Rainforests contain tall trees whose crowns almost touch to form a closed canopy.

Only small amounts of light can filter to the ground. Beneath the tallest trees there are other trees, whose growth is held back because they receive little light. When a tall tree dies, sunlight shines through the gap in the canopy and smaller trees have their chance to grow up to the light.

Tropical and subtropical rainforests can be found in areas of high rainfall (more than 1300 mm per year). They grow best in warmer, sheltered areas on rich soils.

TROPICAL RAINFORESTS

Tropical rainforests are the oldest sorts of rainforest and the most complicated in structure. They have been divided into 13 different types. These vary in plant numbers and species, according to the height of the land and the direction it faces, the rainfall, and the soil. Australia's largest single area (nearly 800 000 ha) is between Cooktown and Townsville, in the Wet Tropics of Queensland World Heritage Area. This covers only 0.001 (one-thousandth) of the area of Australia, but it is home to more than 3000 species of plant, more than 370 species of bird, 30% of the continent's marsupial species, 30% of the frog species and 23% of the reptile species, as well as huge numbers of invertebrates[G].

Fan Palms in tropical rainforest (north-east Q).

SUBTROPICAL RAINFORESTS

Subtropical rainforest in north-eastern NSW.

Subtropical rainforests are like the forests that dominated the world 55–40 million years ago. They are simpler in structure than tropical rainforests and have less diverse plants and animals. They may grow from sea level up to 900 m, favour sheltered situations such as gullies and grow best on richer soils. Today, most of these forests have been cleared. Only small areas remain in south-eastern Queensland and northern New South Wales. Logging, weeds, fire, feral animals and climatic changes all threaten subtropical rainforest.

FACTS 'N' FIGURES FILE

- **Fifty million years ago,** most of Australia was covered by rainforest.
- **Tropical rainforests** contain about 50% of the world's plant and animal species.
- **Every minute, somewhere in the world** an area of rainforest the size of a football field is being felled. This deprives the atmosphere of water and oxygen.
- **Aboriginal people** have lived in NE Queensland rainforests for 40 000 years.
- **Indigenous Australians** probably reduced the size of rainforests by burning the country around them. Fire-resistant eucalypts replaced rainforest trees.
- **Rainforest today covers only about 0.3%** of Australia. Many areas are just patches in gullies, along creeks and in places bushfires cannot reach.
- **Before European settlement,** Australia had perhaps 8 000 000 ha of rainforest. Today, only 2 000 000 ha are left, mostly in isolated patches.
- **Australian's remaining rainforests** would fit into a circle about 160 km in diameter.

In a tropical rainforest

Many different tree species push their crowns into the rainforest canopy.

Up to four layers of plants rise from rainforest floor to canopy.

Rainforest leaves end in drip-tips that shed water quickly.

"Tree gardens" of epiphytes[G] such as ferns and orchids grow on trees.

Palms are common in warm, wet rainforest.

Plank (trunk) buttresses[G] and buttress roots support and anchor tall rainforest trees. The trees' straight, almost unbranched trunks grow upwards towards the light.

Large, woody stemmed vines (lianes) climb up trees to the canopy.

Ferns, lichens and mosses are common on the ground and on the trees.

The leaf litter is thin and is quickly broken down and recycled.

Rainforest plants

SURVIVAL TECHNIQUES

The rainforest is like a battlefield on which plants and animals struggle to survive.

Rainforest plants have had millions of years to perfect ways to defend themselves against insects and other plant-eaters. They have developed ways to hold their own against other plants competing for their space and sunlight.

Here are some of their survival strategies.

POISON

Coppery new rainforest leaves may contain poisonous or nasty-tasting chemicals that repel animals.

ADVERTISE

Rainforest trees that are pollinated[G] by animals bear scented, nectar[G]-filled flowers in easy-to-reach places.

STICK TIGHT

Creepers and vines cling by tendrils, suckers or fine roots, or twine around a supporting tree.

STING

Gympie Stinger leaves are covered with poisonous hairs.

STAB

Lawyer cane is protected by needle-sharp spikes.

STRANGLE

The strangler fig's roots eventually enclose and kill the tree it grows on. The fig takes the dead tree's place in the canopy.

STARVE

The roots of the Curtain Fig make best use of the tree's feeding area. They starve other plants of sunlight and food.

How rainforests work

Rainforests are clean, green machines. They need plenty of rain to grow, but they recycle[G] the moisture endlessly. Three quarters of the rain falling on the closed canopy of a rainforest is sucked up by plant roots. It is then transpired[G] into the air again as water vapour[G].

This water vapour hangs over the forest as mist or cloud, shielding the canopy from the Sun's rays and protecting it from drying out. Eventually, the moist air cools and the water vapour droplets come together into larger drops. When they are heavy enough, these drops fall as rain.

The mist or cloud that hangs over rainforest is made up of water vapour transpired by the forest's leaves.

RAINFOREST RECYCLING

Rainforest ecosystems have rapid nutrient[G] recycling. Organic[G] material that falls on the forest floor, such as leaves and animal bodies, is quickly broken down by fungi, bacteria, worms and other organisms. The nutrients released are sucked up by a network of roots in the top layer of the soil. So food materials are recycled directly into living things, instead of being washed away by the high rainfall.

Most of the food materials in a tropical rainforest system are held in living things and not in the soil. Cutting down a rainforest leaves food-poor soil behind, since plant nutrients disappear with the forest.

FACTS 'N' FIGURES FILE

- **Charles Darwin was a British naturalist** who, in the 1830s, sailed around the world in HMS *Beagle*. He noted that the water vapour that rose from rainforested hills looked like smoke.
- **Plant leaves contain green chlorophyll,** which captures the energy from sunlight. This energy is used to make food using carbon dioxide from the air, and nitrogen and water taken in by the plant's roots. This process is called photosynthesis.
- **The gas oxygen** is a by-product of photosynthesis. Green plants produce the oxygen in the Earth's atmosphere.
- **A rainforest tree may not flower** until it is 30 or 40 years old. Then it may only blossom every three or four years.
- **A patch of rainforest** cannot survive if too many trees are cut from it, because the protective cloud cover they helped to create will be lost.
- **By the year 2050,** eight out of every ten of the world's population will be living in the countries where tropical rainforest survives. Saving some of the forests from clearing will be a huge task.

LIFE AND DEATH IN THE LITTER

Great Barred Frog hunting for food on the leaf litter.

Rainforest leaf litter is rarely deep, because organic material is broken down quickly by fungi, bacteria and small leaf-eating animals. It is the home of many small invertebrates such as worms, slugs and insects. These attract larger predators[G], such as spiders, frogs, snakes, mammals and birds.

A scene in tropical rainforest in Daintree National Park, in Queensland's Wet Tropics World Heritage Area.

PALMS

Palms belong to the group of plants called monocotyledons (= one seed-leaf) that also includes the grasses, orchids and lilies.

Australia has nearly 60 species of palms, about half of them found in rainforest. Their leaves grow directly from the trunk and resemble either a feather or a fan in structure.

Some rainforest palms, such as Lawyer Cane, Wait-a-While and Vicious Hairy Mary, are cane-like, and climb with the help of spines or hooks.

Livistona palms grow near water in desert gorges. They have survived there from long ago when rainforest covered the land.

Palm seeds are held in fleshy fruits. Mammals and birds, such as flying-foxes, possums, fruit-pigeons and cassowaries, eat these. Their bodies digest the fruit and pass out the seeds.

Livistona palms in an inland gorge.

Warm rainforest habitats

THE FOREST FLOOR

The forest floor is often gloomy, for the canopy cuts off light. Plants need light to grow – where a tree has fallen creating a light gap and on the edges of the forest, plants will grow strongly.

Ground-feeding rainforest mammals eat fruits, leaves and fungi. Some animals shelter in the forest but feed near its edges on grassy patches.

The Southern Cassowary, Australia's heaviest bird, feeds on fallen rainforest fruits. It spreads seeds through the forest in its droppings.

The constantly warm, moist climate of tropical rainforests has allowed some ground-living invertebrates, such as earthworms, spiders and centipedes, to become very large.

GROUND-FLOOR RESTAURANT

Trapdoor and funnel-web spiders are common on the floors of warm rainforests. These large spiders live in burrows or silken tubes. They are active at night. Their bite is poisonous.

A dry surface holds light, a wet surface reflects it. Rainforest plants need all the light they can get. Drip-tips help leaves dry quickly so their surfaces take in more light.

The cassowary gulps fruits down. Later it deposits droppings, in which each seed is surrounded by a dollop of fertiliser.

The Musky Rat-kangaroo is the smallest of the kangaroo group. It eats some fruit and seeds, then buries the rest in the leaf litter. If it does not retrieve them, they will grow.

The Cunjevoi Lily grows on the rainforest floor.

Fungi help break down dead plant and animal material.

The Southern Cassowary is an endangered[G], flightless bird.

Red-necked Pademelons graze outside the edges of the forest.

The Buff-breasted Paradise Kingfisher nests in termite[G] mounds on the rainforest floor. It brings insects, snails and frogs to its chicks.

Giant earthworms help recycle the rainforest leaf litter back into soil.

Warm rainforest habitats

TREE TRUNKS, VINES AND EPIPHYTES

Tree trunks and branches serve as highways, stairways, homes and restaurants for animals ranging from possums and tree-kangaroos to lizards, snakes, frogs, snails and insects.

Many animals have regular routes from shelter areas to feeding places and back again. They may travel on climbing plants that grow from the forest floor, or on vines that grow up trees and loop from trunk to trunk.

Epiphytes form habitats in themselves. Their leaves and roots trap moisture and debris, building up a mat on the host tree. Epiphytes and tree-trunk hollows form homes for creatures as different as ants, possums and pythons.

The Leaf-tailed Gecko is a lichen-coloured trunk-dweller.

The Southern Forest Dragon hunts small creatures on trunks and branches.

GOING UP!

Plank buttresses support the trunks of warm rainforest trees that grow in shallow soils. Stretching from the trunk down to the ground, they may send feeding rootlets into the soil.

Ant-plants are epiphytes with swollen stems that house ants. Ant droppings and stored food help feed the plant.

"Highways" regularly used by animals travelling along trunks and branches are kept free of epiphytes and lichens.

Tree-kangaroos climb with strong arms, sharp claws and rough-soled feet. They do not hop but walk along branches.

Vines may be the oldest plants in an area of forest. They survive the death of the trees supporting them, and climb again to the canopy.

The Striped Possum uses a long finger to hook insects from holes in trunks.

Lumholtz's Tree-kangaroo climbs trunks and branches to eat leaves.

The Diamond Python sleeps and hunts in epiphytes and on branches.

Tree-frogs breed in water-filled hollows in trunks and branches.

A Crows Nest Fern may be home to many creatures.

Warm rainforest habitats

THE FOREST CANOPY

The rainforest canopy is made up of the flat, spreading crowns of the tallest trees.

A tree crown may be huge. It is supported on branches that look like the spokes on an umbrella. The bunches of leaves are positioned so that each leaf gets as much sunlight as possible.

There is always a gap of about 1 m between the edges of individual tree crowns (this is known as "crown shyness"). Occasional extra-tall trees, known as emergents, poke up above the canopy.

Canopy flowers may rely on insects, mammals and birds to carry their pollen to other flowers because there is little wind through the treetops. Seeds are often produced in great numbers.

UP IN THE PENTHOUSE

Seedlings of rainforest trees can spend years as spindly saplings. Once a light gap appears, a sapling grows towards the sunlight with great speed.

Some of the saplings that have waited for up to 50 years for a gap in the canopy may be burnt to death by the rays of the Sun.

Predators in Australia's rainforest canopy include snakes, birds of prey and owls.

Cauliflory (producing flowers from the trunk) is seen in lower canopy trees. The flowers and fruits are easily reached by possums, bats and birds.

Some birds take only 5 min to digest rainforest fruit and pass the seeds.

The Red-eyed Tree-frog lives in the canopy. It breeds in ground pools.

The Green Ringtail Possum sleeps on a branch. It eats canopy leaves.

Birdwing butterflies feed on nectar when rainforest trees blossom.

The Green Tree-snake catches frogs and birds in the canopy.

Flying-foxes eat nectar and soft fruits from canopy trees.

The Little Shrike-thrush feeds on insects that it catches in the canopy.

The Wompoo Pigeon feeds on fruits in the canopy.

Warm rainforest habitats

STREAMS, RIVERS AND WATERFALLS

Rain falling on a rainforest runs across the forest floor, gathering in trickles and creeklets that join up to form creeks and rivers.

Some of the water that soaks into the ground seeps into these waterways. Where the forest is on steep slopes, the streams flow fast. When they flow over harder rocks, a waterfall is formed.

Rainforest waterways and their banks are homes to insects, crustaceans[G], worms and other invertebrates that are eaten by predators such as fish, turtles and the Platypus. Water animals form food for birds, such as herons, and for reptiles, such as monitors and dragon lizards.

Waterways may disperse rainforest seeds.

The Eastern Water Dragon is at home in water or on land. It eats plant matter as well as tadpoles and other water creatures.

The Lace-lid Frog is found in north Queensland rainforests.

Permanent freshwater pools are habitat for Freshwater Crocodiles.

POOLS, SPAS, JACUZZIS

Nature's jacuzzi is the plunge-pool formed at the base of a waterfall.

Waterlogged soil does not suit the growth of rainforest. Sloping land and plenty of creeks drain off excess water and keep the forest healthy.

Fast-flowing mountain streams are the habitat of the gastric-brooding frogs, whose tadpoles mature[G] in their mother's stomach. These and some other mountain stream frogs may be extinct[G].

Freshwater crays are common in rainforest streams. The moist forest atmosphere allows these crustaceans to spend time out of water.

Freshwater Crocodiles eat fish and other creatures. They do not eat humans.

The Gastric-brooding Frog once lived in highland forest streams.

The Lamington Blue Cray sometimes travels across the damp forest floor.

The Nankeen Night Heron hunts in rainforest creeks.

Monsoon rainforests

A monsoon[G] forest is watered by heavy rains from summertime's tropical monsoons. The forest then must survive a period in which almost no rain falls. In Cape York Peninsula, Q, the Top End, NT, and the Kimberley, WA, monsoon forests are found in scattered patches.

Some monsoon forests grow on rocky places. Their trees, tangled vines and shrubs may lose their leaves in the dry season. Other monsoon forests grow around spring-fed creeks and rivers. Their tall evergreen trees and palms may grow over 25 m in height and do not lose their leaves in the Dry.

Monsoon rainforests are the remains of rainforests that grew across Australia as recently as 15 million years ago. Often these patches of forest are small. Patches around water may be as small as 1 ha in area, while those on drier soil may be as large as 100 ha. The rainfall, over 1200 mm per year in most places, comes in summer and it is hot to very hot all year around.

These rainforests have to deal with extremes of heat combined with flooding rain at some times of year and drought at other times.

Palms, pandanus, cycads, tall grasses and tangled vines grow in the understorey of these forests.

Monsoon rainforest near water in the Top End, NT.

The Pig-nosed Turtle eats fruit that has fallen into the stream it lives in.

Mertens's Water Monitor hunts tadpoles, fish and other water life.

The Agile Wallaby spends the day in the shelter of the monsoon forest.

FACTS 'N' FIGURES FILE

- **In the late dry season,** when temperatures soar, monsoon rainforests provide shelter for animals that usually live in other habitats.
- **The springs that water** monsoon forests in valleys and under cliffs provide oases at which animals such as wallabies can drink.
- **Many national parks** in the Top End and Kimberley have tourist attractions that feature waterfalls and monsoon forest.
- **In the Kimberley** of Western Australia, many patches of monsoon rainforest have their own unique species of native snail.
- **Cattle and pigs** destroy monsoon rainforest by wallowing in the wet, green areas around springs and creeks.
- **Monsoon rainforest trees** may flower in the dry months of the year.
- **Some stands of monsoon forest** growing along stream banks may be several kilometres long but only 100 m wide.
- **Frequent fires** have cut back the areas of monsoon forest in northern Australia. Forests on rocky hills are vulnerable[G].

The monsoon rainforest shown above surrounds the plunge pool formed by Florence Falls in Litchfield National Park, in the Top End of the Northern Territory. Inset: Tree-snakes live in this forest. They eat birds, reptiles and small mammals.

Temperate rainforests

Above: Cool temperate rainforest, Maryville, V. Above right: Platypus.

Above: Cool temperate rainforest, Mt Field National Park, T.
Left: Tasmanian Devils.

Temperate rainforests

Cool temperate rainforests contain some of the most ancient species of Australia's plants.

Many of the ancestors of these plants grew in Gondwana over 60 million years ago. Today, these forests share plant groups with similar forests in Africa, South America and New Zealand – areas that were also part of Gondwana.

Fossils of cool temperate rainforest trees have been retrieved from under the icecap that today covers Antarctica, also once part of Gondwana.

COOL TEMPERATE RAINFORESTS

Cool temperate rainforests grow in areas of very high rainfall (over 1750 mm per year) and usually high above sea level (over 900 m). There may be two layers of trees, but sometimes only one. This sort of forest contains fewer tree species than warmer rainforests and the outstanding tree is often the Antarctic Beech. There are usually ground ferns and tree ferns. The trees support mossy epiphytes, lichens and thin, wiry vines.

In **New South Wales**, large areas of cool temperate rainforest have been cleared. The tallest timber has been logged from many remaining stands. **Victoria** has four major areas of cool temperate rainforest left after intensive clearing. **Tasmania** has Australia's largest areas of cool temperate rainforest, covering about 10% of the State.

Cool temperate rainforest in Victoria.

FACTS 'N' FIGURES FILE

- **A Southern (Antarctic) Beech tree** may survive for more than 500 years and grow to over 50 m in height.
- **Two-thirds of Tasmania's** cool temperate rainforest is not protected as parkland. It is under threat from logging, mining, tourist roads and facilities, and fire.
- **Fifty per cent of the rainforest** and tall eucalyptus forest that existed in Tasmania 200 years ago has been cleared.
- **More than 75%** of the rainforest that existed in New South Wales 200 years ago has been cleared.
- **Nearly 80 of Victoria's** rainforest plant species are rare or threatened with extinction.
- **Cool temperate rainforests** are often hidden in mists when it is not raining.
- **Cool temperate rainforest** existed well before eucalypts, acacias and other drought-resistant plants developed.
- **A mixed forest** is one where rainforest grows beneath the shelter of tall eucalypts.
- **The greatest threats** to cool temperate rainforest come from human activities, including fires, clearing, and the pests and diseases that enter the forests after disturbance by humans.
- **Bushfires that spread** from neighbouring eucalypt forests destroy rainforest trees, which do not sprout again as many eucalypts do.
- **Phytophthora root-rot** may kill rainforest trees that are regrowing on recently burnt land.

The dominant tree species of cool temperate rainforest is the Southern (Antarctic) Beech.

In a temperate rainforest

There are usually two layers of plants, but sometimes only one.

Cool temperate rainforests are often cloaked in mist.

Most temperate rainforest plants are evergreens that always have leaf cover.

Ground ferns and tree ferns are common in cool temperate rainforest. They form an understorey and provide places for many small animals to live.

If the forest has not been logged, the tree trunks may be of massive size.

CONSPICUOUS BY THEIR ABSENCE

Some things very noticeable in tropical and subtropical rainforests are absent from cool temperate forests, including:

- strangler figs ✘
- palm trees ✘
- plank buttresses ✘
- large vines (lianes) ✘
- large epiphytes (ferns, orchids etc.) ✘

PRESENT AND CORRECT

Some of the things present in cool temperate forests include:

- Southern (Antarctic) Beech trees ✔
- thin wiry vines ✔
- some small epiphytic orchids ✔
- abundant mossy epiphytes ✔
- lichens ✔
- ground ferns and tree ferns ✔

Lichens and mosses are common on trunks, branches, rocks and logs.

Types of growing things

FERNS

Ferns grow best in damp places. They reproduce by forming spores on the undersides of leaves. A spore grows into a tiny, heart-shaped plant that carries male and female organs. These produce sperm and eggs that join and grow into a fern. The ancestors of ferns helped form Earth's coal reserves.

FUNGI

The body of a fungus, called a mycelium, usually grows as a mass of hair-like structures. A fungus reproduces by spores that grow on fruiting bodies, e.g. mushrooms and toadstools. Fungi are able to break down substances by oozing digestive juices onto them, then sucking up the liquid. Fungi do not need sunlight.

LICHENS

A lichen is a partnership between a fungus and a tiny organism that lives in the fungus's mycelium and carries out photosynthesis. The fungus absorbs minerals and water, while the partner captures light energy to produce food. Lichens can survive in extreme climatic conditions. They are important nutrient recyclers.

PANDANI – A FLOWERING PLANT

Found only in Tasmania, the Pandani is the largest heath plant in the world. It is not related to the tropical pandanus palms. The Pandani grows in cool temperate rainforest and also on mountains. The long, sharp-edged leaves remain on the trunk and provide protection from very low temperatures.

Temperate rainforest animals

Tasmanian rainforest has a smaller variety of animals than live in mainland forests.

Animals in these southern forests have to survive very cold winters. The mammals, such as possums and wallabies, grow thick fur. Smaller species become torpid[G] during very cold spells.

Reptiles go into winter sleep, or hibernation, and emerge only on warm days. Many birds leave the forests in winter. They fly to warmer coastal areas, returning to nest in spring and summer.

Many of Tasmania's rainforest mammals come out to feed after dark. Since foxes have not reached the island, these include small ground-living species that are now uncommon on the mainland.

Tree fern.

The Sugar Glider swoops between trees on skin stretching between fore and hind limbs. Members of a colony sleep in a tree hollow. They recognise other group members by scent.

The Spotted-tailed Quoll is a predator on small mammals and birds. It is Australia's second-largest carnivorous marsupial (the Tasmanian Devil is the largest).

Temperate rainforest animals

Victorian rainforest is very much reduced in area by logging, clearing and fires.

There are now only four major areas of rainforest left in all of the State. Each one is made up of many narrow stands, the largest of these stands only about 10 ha in area.

Gippsland's rainforest was extensively cleared during the late 1800s and there is not much left of the thick forests that covered the area before European settlement. Small areas of rainforest remain in the Strzelecki Ranges, especially in the east on steeper, higher slopes. Other areas are in the Central Highlands and in the Otway Ranges on the central western coast.

Groups of Feathertail Gliders sleep in tree hollows. They eat pollen, nectar and insects.

The Common Brushtail Possum eats fruits and leaves. It spends the daytime in a tree hollow.

Rainforest stream.

The day-active, frog-eating Red-bellied Black Snake may be seen along stream edges.

The Long-nosed Potoroo digs up and eats the underground fruits of fungi.

Sclerophyll forests

Above: Wet sclerophyll forest in Victoria. The trees are Mountain Ash, growing above an understorey of ferns.
Inset: The Yellow-bellied Glider eats eucalypt sap, nectar and pollen.

Above: Dry sclerophyll forest in the south-west of Western Australia. The trees are Wandoo.
Inset: The Purple-crowned Lorikeet eats nectar from eucalypt flowers.

Sclerophyll forests

Sclerophyll forests are open canopy forests consisting of plants seen as typically Australian, such as eucalypts, tea-trees, banksias and wattles.

The word sclerophyll comes from *sclero* = hard and *phyllon* = a leaf, and these plants have tough leaves, often spiny or very small. These slow down the rate at which moisture is lost from the plant in dry conditions. Many Australian soils are low in the nutrient phosphorus, and sclerophyllous plants have evolved to grow well in these soils.

Sometimes they are helped by fungi that grow on their roots and fix nitrogen from the soil.

WET SCLEROPHYLL FORESTS

The tallest trees in a wet sclerophyll forest may be over 30 m in height. The understorey plants, such as ferns, tree ferns, creepers and shrubs, have soft leaves. Australia has 6 500 000 ha of wet sclerophyll forests (4% of total native forests), most in Victoria, New South Wales, Queensland and Tasmania. These forests may contain very tall trees, such as the Mountain Ash (Victoria and Tasmania) and the Karri and Tingles (Western Australia). Other trees may include Mountain Grey Gum and Tasmanian Blue Gum.

Wet sclerophyll forest, south-west WA.

DRY SCLEROPHYLL FORESTS

Dry sclerophyll forest, New England, NSW.

The trees that grow in a dry sclerophyll forest are 10–30 m tall. The plants that grow beneath them have tough, hard leaves. Dry sclerophyll forests contain many different plant species, although the dominant trees are usually eucalypts. They also contain many species found nowhere but in that region. These forests make up more than half of Australia's total native forest, and occur in every State and Territory. Medium height sclerophyll forest shades into woodland, and some sorts of plants grow in both habitats.

FACTS 'N' FIGURES FILE

- **Eucalypts dominate** 124 000 000 ha of Australia's open forests and woodlands.
- **A River Red Gum** near Mt Gambier, SA, is thought to be at least 800 years old.
- **Near Sydney,** a patch of dry sclerophyll forest 0.1 ha in size could contain 70–100 different sorts of plants.
- **In 1642, Dutch captain Abel Tasman** listed plants his crew found at Blackman Bay, Tasmania. They included "good gum which is dripping from trees", probably sap from stringybark, a type of eucalypt.
- **Some eucalypt species** take different forms according to the environments they grow in. So a eucalypt tree might have one thick trunk in a well-watered climate, but several slender trunks in drier areas.
- **Seventy-five per cent** of plant species in the dry sclerophyll forests of the south-west of Western Australia are found nowhere else in the world.
- **Sclerophyllous plants** often interact with animals. While feeding on nectar, the animals carry pollen between flowers.

Sclerophyllous plants

Sclerophyllous plants have hard, often spiky, waxy or hairy leaves. The hardness in the leaves comes from lignin, which prevents the leaves wilting in dry conditions.

Sclerophyllous plants can grow in soils with low phosphorous content. They include eucalypts, tea-trees and bottlebrushes, banksias and grevilleas, wattles, and peas.

FAMILY: MYRTACEAE

Woody shrubs and trees with simple[G] leaves dotted with glands[G] containing aromatic[G] oil. Flowers are wheel-like when viewed from above. There are numerous pollen-bearing stamens. The fruit may be a woody capsule or a fleshy berry.

Eucalypt buds, flowers and leaves.

Bottlebrush flowers and leaves.

FAMILY: PROTEACEAE

Shrubs or trees with leathery or stiff leaves, often with saw-like or indented edges. New growth may be covered with bronze hairs. Flowers appear as cones, in spikes or in dense heads, and are often rich in nectar.

Banksia flower cone and leaves.

Grevillea flower spikes.

FAMILY: MIMOSACEAE

Woody trees or shrubs. Seedling leaves are divided and feathery, while adult foliage consists of flat phyllodes[G]. Bark may be rich in tannins[G]. Flowers are small, fluffy and yellow. Fruit is a pod containing hard seeds.

Wattle flowers and phyllodes.

FAMILY: FABACEAE

Shrubs or herbs; many climb by tendrils. Leaves often compound[G]; flower has 5 petals; fruit is pea-like and enclosed in a pod. Nodules[G] on roots contain bacteria that help plants take in nitrogen.

Pea flowers and leaves.

Eucalypts

AND THEIR ALLIES

Eucalypt trees are often called "gum trees" because of their gum-like sap.

In Australia they can be found from the snowline of the highest mountains to the desert inland. Rainforest is the only major environment where there are no eucalypts.

Australia has around 700 species of eucalypts and eucalypt relatives. Twelve more species occur naturally outside Australia and eucalypts have been planted in many overseas countries.

Botanists[G] classify eucalypts into three groups: eucalypts, corymbias and angophoras.

EUCALYPTS

- More than 500 species.
- Adult leaves come off the stem alternately.
- Flower has a cap (operculum) that falls off as the flower opens.
- Flower heads have flowers at many different levels.

CORYMBIAS

- 113 species.
- Adult leaves come off the stem alternately (like those of eucalypts).
- Flower has an operculum.
- Flowers are in corymbs. Their stalks arise from different levels on the stem, but all the flowers are level.

ANGOPHORAS

- 13 species.
- Adult leaves come off the stem opposite each other.
- Flower does not have an operculum (unlike other two groups).
- Flowers are in corymbs. They have different length stalks but end up level.

FLOWERS, FRUITS AND BARK OF SOME EUCALYPTS

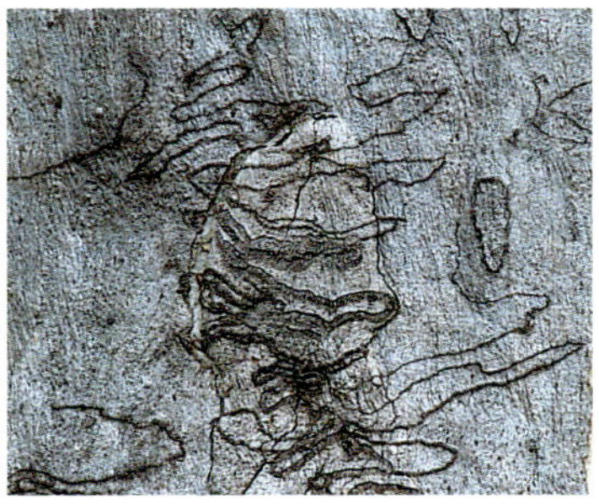

WHAT'S IN A NAME?

Ash timber reminded early settlers of the European Ash.
Beefwood has reddish wood.
Box has bark that is shed in plates or pieces.
Blackbutt trunks are covered by fire-darkened bark.
Bloodwood timber has pockets of dark red gum.
Ghost gum has ghostly white bark.
Gum has smooth bark that may ooze sap.
Ironbark has very hard, deeply furrowed bark.
Mallee trees have many stems arising from a large root.
Peppermint Gum has peppermint-scented leaves.
Ribbon gum bark is shed in long strips.
River Red Gum wood is rich red when freshly cut.
Salmon Gum has pink bark, like Atlantic salmon flesh.
Scribbly Gum bark is marked by insect tunnels.
Stringybark has bark that comes away in long fibres.
Woollybutt has rough, woolly bark.

Wet sclerophyll habitats

THE FOREST FLOOR

The wet eucalypt forests of the high rainfall regions of south-eastern and south-western Australia have very tall, old trees. These tower above dense understoreys of shrubs, tree ferns, creepers and climbing plants.

These forests grow on deep soil in areas that receive high rainfall.

When a mature forest is destroyed by severe fires, new trees grow from seeds carried in capsules that only open after being scorched by flames. This means that all the canopy trees of a wet eucalypt forest are usually much the same height.

The new canopy trees prevent smaller trees growing up beneath them.

TALL TIMBERS, NATURAL TREASURES

Old-growth eucalypt forests are under threat and increasingly rare. They are the last refuges of some endangered creatures.

The Mountain Ash is the tallest hardwood in the world (an 1872 record was 150 m, the modern record is 95 m).

A Karri tree in Warren National Park, WA, is 89 m tall. The Gloucester Tree, near Pemberton, WA, holds the highest (60 m) fire lookout in the world.

Small areas of wet sclerophyll forest occur in north-eastern Queensland on the western edge of rainforest above 600 m. Fire prevents rainforest from invading these forests.

The Common Wombat burrows and grazes around forest edges.

Superb Lyrebirds scratch their insect food from the forest litter.

Antechinus and other small marsupials hunt the forest floor.

Plants that grow in the understorey of Western Australia's Karri forests include ferns, grasstrees, wattles and many other species.

The Water-rat lives around creeks and rivers, eating aquatic animals.

Wet sclerophyll habitats

TREE TRUNKS, BRANCHES AND LEAVES

The Koala wedges itself into a tree fork to sleep during the day. However, the other furred animals of Australia's open forests need somewhere to hide from the sun, the weather and predators.

Tree hollows are the preferred home, but hardwood trees such as eucalypts take many years to develop suitable holes. Unless considerable areas of old forest are left unlogged, many tree-living animals will have no homes.

Even in a hollow, a possum, glider or parrot can be at risk from one of Australia's larger reptile predators, the climbing Lace Monitor. At night, Masked and Powerful Owls hunt unwary treetop animals.

The Greater Glider eats mostly eucalypt leaves.

The Crimson Rosella nests in tree hollows and eats fruit and seeds.

HOLLOW SLEEPERS

Leadbeater's Possum was thought to be extinct, but was rediscovered in 1961. It eats insects and spiders, eucalypt sap and honeydew formed by sap-sucking insects. Only 3% of its habitat is in protected areas.

Hollows suitable as homes for the rare Leadbeater's Possum do not develop until a tree is at least 120 years old and at least 1 m in diameter.

A Mountain Grey Gum does not develop hollows until it is over 700 years old.

The Greater Glider eats almost nothing but eucalypt leaves. It can glide up to 100 m on membranes[G] stretching between the hind foot and the elbow on each side.

This Leadbeater's Possum is licking sap from a cut in tree bark.

The Lace Monitor climbs trees to hunt for birds' eggs and nestlings.

The Masked Owl takes prey from the branches.

Dry sclerophyll habitats

THE FOREST FLOOR

The floor of a dry sclerophyll forest consists of open spaces covered with leaf and bark litter, and with bushes and herbs.

Often the ground storey plants, which include orchids, pea flowers, wattles and grevilleas, flower plentifully in spring and early summer.

The animals that feed on the floor of this forest may eat leaves and dig up roots and underground fungi. They may hunt invertebrates in the leaf litter and fallen logs.

If ground litter builds up, hot fires can sweep through these forests. Many of the plants have some way of regrowth, or have seeds that survive and sprout after fire.

A TERMITE CONNOISSEUR

The Numbat was once found in many forest areas across southern Australia. Today only a few small groups of Numbats remain. The largest is in Wandoo forest near Dryandra, Western Australia.

This day-active marsupial has small teeth and a very long, sticky tongue. It rakes open surface termite runways and licks up the insects. A female Numbat has up to four babies. She carries them on her teats for some months, then they are left in a nursery burrow. Numbats are in danger from foxes, bushfires and forest clearing. Their numbers build up after foxes are removed from an area.

The Numbat is one of the few marsupials that is active during the day. At night it sleeps in a hollow log on or near the forest floor.

Short-beaked Echidnas roam the forest floor. They lick up termites and ants with their sticky tongues.

Bettongs are small kangaroo relatives. They feed on plants, insects and underground fungi.

Dry sclerophyll habitats

TREE TRUNKS, BRANCHES AND LEAVES

The trunks and branches of trees in a dry sclerophyll forest develop hollows that are used by many creatures as homes and nurseries.

Many birds, such as treecreepers and sittellas, find their food on the tree trunks, and a host of small birds pick leaf-eating insects from the foliage.

The fruits of dry sclerophyll forest trees are often protected by a woody fruit capsule, such as a gumnut. Some parrots specialise in cracking open these capsules and eating the seeds.

If fire in dry sclerophyll forest catches the canopy, it is fuelled by leaf oils and may travel swiftly through the treetops. The trees will grow again from their roots, and from buds protected by bark.

TREETOPS GOURMET

The Koala is adapted to a lifetime spent amongst the branches. It has long, strong limbs ending in sharp-clawed, grasping paws (each "hand" has two "thumbs"). Its fur is waterproof, and during the day it sleeps propped in a tree fork. At night it climbs around eating eucalypt leaves. Its gut organisms break down the leaf fibre and its liver detoxifies[G] the poisons in the leaves. However, a Koala gets little energy from its food and sleeps up to 20 h out of every 24. A baby Koala is carried in its mother's pouch for 6 months, then rides on her back until it is around one year of age.

The Koala's diet consists mainly of eucalypt leaves. Tiny organisms in the Koala's very long gut break down the tough leaves so they can be digested.

This Eastern Quoll and her young eat birds, frogs, insects and small mammals.

Rufous Treecreepers cling to tree trunks, hunting for insects.

Yellow-tailed Black-Cockatoos use their strong beaks to get seeds.

Callitris, acacia and casuarina

Trunk of a callitris, or cypress pine.

CALLITRIS FOREST

Callitris are Australian cypress pines, which are survivors from the days of Gondwana. Their bark has deep furrows and extends to the smallest branches. Their leaves are tiny and scale-like. White cypress, the tree most important to humans, has a straight trunk and can grow to 30 m, but is usually 15–20 m. Callitris forests are drought-resistant and are usually found in areas with a rainfall of 300–650 mm a year. They are most common in eastern Australia where only five major areas remain. There are around 1 000 000 ha of callitris forest. Cypress pine is termite-resistant and is used in housing construction and for flooring, walls and fence posts. Large areas of callitris are managed for long-term timber production.

ACACIA FOREST

Acacias are usually called wattles. They occur on all continents except Antarctica, mostly in warmer regions. There are over 700 Australian species, all over the country. In wetter areas they form tall, open forests. In drier regions they form low woodlands. Brigalow, lancewoods, gidgees, mulgas, myalls and blackwoods are all acacias, and they make up the second most extensive of Australia's forest types after eucalypts. There are 12 000 000 ha of acacia forest (8% of the total native forest area). Nitrogen-fixing bacteria living in nodules on the roots of acacias put nutrients into the soil they grow in. Although most acacias live less than 20 years, mulga, blackwood and cedar wattle may live to more than 100 years. Wattles supply leather-tanning chemicals, dyes and fence posts. Cattle and sheep eat their leaves. The Aborigines ate wattle seeds and gum, and used the wood for spears and the bark for sandals.

A Long-billed Corella separating acacia seeds from their pods.

Casuarinas grow along watercourses.

CASUARINA FOREST

Casuarinas are also known as she-oaks. Their leaves look like those of pine trees, but the drooping needles are actually branchlets ringed with tiny leaves, only the tips visible to the naked eye. These branchlets can carry out photosynthesis. There are around 60 Australian casuarina species, including Belah, River Oak and Swamp Oak. Casuarinas are found from southern Queensland across to central southern Western Australia. River Oak may grow to 35 m in height along watercourses, replaced by Swamp Oak in brackish coastal swamps. There are about 1 000 000 ha of casuarina forest in Australia, mostly in New South Wales. West Australian She-oak is used in the furniture industry. Belah is used for fence posts.

Melaleuca and mangrove

Common Ringtail Possum in melaleuca.

MELALEUCA FORESTS

Melaleuca species are usually known as paperbarks because their bark peels away in thin, papery layers of cork. Melaleucas usually grow in damp or wet country, or in coastal areas that dry out seasonally. Often melaleuca forests are narrow strips of trees along streams and swamps. In northern Australia they are spread across swamps that fill in the summer Wet. The leaves of some melaleuca species supply tea-tree oil, which is used as an antiseptic, or manufactured into creams, shampoos, soaps, insect repellents, perfumes and flavourings. There are around 4 000 000 ha of melaleuca forests in Australia. Over 85% of them are on privately owned or leased[G] land, which is generally used for cattle production. Only 10% of Australia's melaleuca forests are protected in conservation reserves.

MANGROVE FORESTS

Mangroves grow on mud and sand in sheltered estuaries[G], inlets and bays that are flooded by saltwater tides. Mangroves have shallow, spreading roots that send up aerating roots: the aerating roots take in oxygen when the parent roots are flooded. Mangrove trees get rid of excess salt by storing it in leaves that are then shed. In northern Australia, mangrove trees may reach 30 m in height; further south they range from 3 to 8 m high. They hold the shoreline in place, provide habitat and breeding sites for birds and other wildlife, and are vital nurseries for many fish, crab and mollusc species. Around 40% of mangrove forests are privately owned, 12% are on leased land and 22% are in conservation reserves.

Mangrove forest along the Daintree River.

FACTS 'N' FIGURES FILE

- **The Pilliga Scrub,** north of Coonabarabran, NSW, is a callitris forest. It is the largest remaining native forest in NSW west of the Great Divide. Other NSW callitris forests have been cleared for agriculture. Small stands remain along roads.
- **"She oak"** was a name given to casuarinas by Europeans because the wood looked like oak wood but was not so strong.
- **Casuarinas** were so-called because it was thought their leaves resembled the feathers of the Southern Cassowary. This bird's Malay name is Kasuari, its scientific name is *Casuarius casuarius*.
- **Acacia bark** was used by Aborigines to poison fish and also to weave nets and bags. The gum was used as an adhesive[G]. Witchetty grubs, which form good bush tucker, live in the roots of some acacias.
- **Aboriginal uses** for melaleucas included materials for canoes, shelters, blankets, burial shrouds and bandages. Flowers were soaked in water to prepare a sweet drink. Leaves were burnt to repel mosquitoes and were used to treat colds, aches and coughs.
- **Mangroves grow best** in warmer climates. Australia has around 35 species, but none grow in Tasmania.

Parasites and predators

MISTLETOES AND MISTLETOEBIRDS

Mistletoes have been part of Australia's flora for at least 70 million years. These plants are parasites[G] that tap into their host tree's water and mineral supply. Some mistletoes live only on one sort of tree, some live on many and some even attack other mistletoes. Mistletoe seeds are packed in sweet flesh that contains sticky threads and is surrounded by a leathery skin. Many birds eat mistletoe nectar and pollinate mistletoe flowers. A number of birds eat mistletoe berries, but the tiny Mistletoebird is a specialist. A meal will pass through a Mistletoebird in 3–12 min, then the bird wipes the seeds dangling from its bottom onto a branch. The sticky threads dry to form a plastic-like shield. Each seedling glues itself to its host and pushes out roots into the host's tissue.

A Mistletoebird chick passes sticky mistletoe seeds.

FACTS 'N' FIGURES FILE

- **Tasmania** has no mistletoes and no Mistletoebirds.
- **Some mistletoes** have leaves that look like the leaves of the host plant.
- **As forests are thinned out,** and more light enters, mistletoes increase in numbers. When a tree's roots have to support more mistletoe leaves than tree leaves, it can die.
- **An Aboriginal legend** says that spirit babies hide in trees until women walk by. The babies unable to find a mother wail for a time, then become mistletoes. The red colour of mistletoe flowers is thought to be like baby's blood.
- **Nectar glands** around the rim of a pitcher plant attract insects. An overhanging inner rim and slippery walls stop them climbing out.

PREDATORY PLANTS

A pitcher plant traps animal food.

Australia has about 70 species of sundew.

Carnivorous plants growing on nutrient-poor soils obtain nitrogen from digesting small creatures. Pitcher plants have special leaves that form smooth-walled traps with digestive juices at the base. Sundew leaves are covered with sensitive, sticky hairs that hold insects. Other hairs secrete[G] juices that digest the prey.

Fungi

HELPFUL FUNGI – FOODMAKERS

A mushroom or toadstool is the fruit of a fungus. The body of the fungus, the mycelium, grows as a mass of thin threads. Fungi can grow in a wide variety of temperatures (–5°C to 50°C). They do not require light. A fungus secretes substances that break down the material surrounding it so it can absorb the digested food. Fungi decompose organic matter and can digest the woody lignin that supports plant tissues. Certain fungi can form a "nutrient bridge" between the soil and plant roots. They take in phosphorus, zinc, copper and other minerals and pass them to the plant, then take processed sugars back again. These fungi can help a plant get 10 000 times more phosphorus than normal, and help it combat heat, drought and disease. Some fungi can be eaten by humans, who use them to flavour cheese and produce biochemicals. Yeasts are fungi used in making wine and bread.

Four different types of fungal fruiting bodies.

FACTS 'N' FIGURES FILE

- **Fungi and bacteria break down** dead plant and animal bodies. Without fungi, these would cover the land.
- **Potoroos, bettongs and bandicoots** dig up and eat the fruiting bodies of forest fungi. They spread the fungus spores in their droppings and play an important role in the forest ecosystems.
- **Soy beans** cannot be digested by humans. However, fungi turn beans into digestible bean curd or soy sauce.
- **Fungi poisonous to humans** include many toadstools and mushrooms, the ergot fungus that infects wheat and rye, and a fungus that forms mould on peanuts and produces toxins[G] that can kill humans.
- **Ringworm** is caused by a fungus.
- **Many poisonous fungus** species look very like harmless mushrooms. "Do not experiment" is a good rule to follow.

HARMFUL FUNGI – DIEBACK

Fungi are important causes of plant diseases. Because modern agriculture farms one variety of one species of plant in one place (monoculture), harmful fungi cause a lot of damage. Eucalypts and other sclerophyllous plants are attacked by many fungi, including one that causes dieback. Foliage is destroyed and a white mat of mycelium grows under the bark of infected roots, killing the tree. Dieback occurs in Australian forests, woodlands and gardens, and is most severe where cars have compacted the topsoil.

Dieback is a fungal disease that can kill trees.

Flowers and fruits

POLLINATION AND DISPERSAL

To reproduce, the male parts of a flowering plant produce pollen grains. These are carried to the female parts of the same plant or another plant of the same species. The pollen fertilises an ovule that develops into a seed containing an embryo[G], which develops into a new plant.

Both pollen and seeds need to be carried away from their parent plants. Forest trees may use wind to carry their pollen and water to disperse their seeds, but often pollen and seeds are carried by animals. Flowers attract these carriers by providing energy-rich nectar and pollen. They advertise their presence with scent, colour and pattern. Seed is often packaged in a fleshy fruit. An animal eats the fruit, digests the flesh and the seed passes out with the animal's droppings, surrounded by fertiliser.

Fruit- and nectar-eating bats pollinate flowers and spread seeds.

New Holland and other honeyeaters eat nectar and carry pollen on their beaks.

A Rainbow Lorikeet uses its brush-tipped tongue to get nectar.

An Australian Ringneck eats the seeds of a eucalypt.

Eastern Pygmy-possums eat pollen and nectar. They carry pollen between flowers.

Sulphur-crested Cockatoos crack open hard fruits to get seeds.

The Musky Rat-kangaroo may bury rainforest seeds, then forget to dig them up.

Beetles, bees, wasps and other insects carry pollen between flowers.

Ecological interactions

The preceding page describes ways in which Australian plants and animals interact.

In many cases, these interactions are unique to Australia. Over long periods of time, the interacting organisms have evolved together. As one changed, so did the other.

Many features of Australia's flowering plants can only be understood by seeing the ways in which they interact with animals. Many animals are built, and behave, in ways that have resulted from their interactions with plants. These interactions include animals pollinating plants and dispersing their fruits and seeds, and the defences that plants have developed against animals that want to eat them.

THE SUGAR GLIDER AND THE EUCALYPT TREE

An interaction that benefits both parties

Sugar Gliders sleep in hollows in mature eucalypt trees. Up to seven adults and their young ones may share a nest.

At night, a Sugar Glider feeds on nectar and is dusted with pollen. This is carried to other blossoms to pollinate them.

Gliders bite holes in bark, then lap up the tree sap that oozes out.

The Sugar Glider eats insects that could damage the tree.

LIVING IN COMMUNITIES

A community is an assembly of species that interact with one another. Various sorts of relationships are possible between two organisms in a community.

- **Mutualism** Both organisms benefit (e.g. Sugar Glider and eucalypt tree).
- **Commensalism** One organism benefits and the other is not affected in any way (e.g. epiphytic plant and rainforest tree).
- **Parasitism** One benefits and one is harmed (e.g. mistletoe and host tree).

Forest food-chains

Forest trees form food for many sorts of animals.

Phasmids (stick insects) eat leaves. They rely on camouflage to protect them.

The Pacific Baza feeds on phasmids and other foliage animals such as tree-frogs.

Cicada nymphs[G] live in the soil, sucking root sap.

Bandicoots eat insects, plants and fungi. Their droppings and bodies feed the forest.

The Little Eagle eats mammals, such as bandicoots and rabbits, as well as birds.

Grass and herbs are fertilised by animal droppings.

Eastern Grey Kangaroos and other grazers create "marsupial lawns".

Dingos hunting in a pack are capable of killing even adult kangaroos.

Food-chains start with plants, which make food using the energy of the Sun.

Herbivores eat plants.

Carnivores eat other animals (**predators** kill them and eat them, **scavengers** eat dead animals).

Omnivores eat everything edible.

Fire and the forests

A CONTINENT OF FIRES

Fire is a natural part of Australia's ecosystems and habitats. It does not occur at regular intervals, or in similar intensities, unless assisted by human actions. Many Australian plants have adapted to particular fire regimes[G], and some depend on fires to open seed pods and fruits, and to provide rich ash for seedlings to grow in.

Rainforest does not burn readily. Its foliage does not catch fire easily, it grows in areas of high rainfall, the air in and around it is moist, and there is little forest litter. If it is burnt, trees that are more fire-resistant, but also more easily burnt, take its place. The rainforest becomes sclerophyll forest.

Eucalypts and other sclerophyllous trees have foliage rich in oils that burn easily. Fire may leap from crown to crown, travelling with great speed across the canopy. Shed leaves and bark create deep litter in sclerophyllous forests, and this contributes to hot, slow-moving fires. Eucalypts grow again after most fires.

Sclerophyll forest burning.

Trunk regeneration[G] of a eucalypt.

New growth after a bushfire.

HOW EUCALYPTS RECOVER FROM FIRE

At, or below, ground level, most eucalypts develop a swollen, woody structure called a **lignotuber**. This contains food reserves and buds. When the above-ground parts of the plant are destroyed, e.g. by fire or logging, these buds develop into new shoots. If only the crown and small branches are killed by fire, buds lying just under the surface of the trunk and branches sprout to form shoots. These die off after the crown has grown again.

A few eucalypts, such as the Mountain Ash, have thin bark and no lignotuber. They are easily killed by fire and must replace themselves from seed. These seeds are protected by capsules that open to release them soon after the fire has died out. Many other sclerophyllous plants, such as banksias, also release their seeds after their woody fruits have been scorched, heated or smoked.

Banksias release seeds after fire.

Humans and forests

When Europeans arrived in Australia in 1788, about 10% of the continent was covered with forest. Since then, about half of these forests have been cleared. This includes 75% of the rainforests and over 90% of the old growth forests. Forests now cover just 5% of Australia's land area.

Around 110 000 ha of state forest are logged each year (the same area as 1 000 000 suburban house blocks). Around 200 000 ha of all publicly owned native forests, including Crown (Government-owned) land, are logged each year – an area almost the size of the Australian Capital Territory.

FORESTS AND THE GREENHOUSE EFFECT

Forests soak up huge amounts of carbon dioxide, a gas, and give out oxygen. If a forest is logged, the increase in carbon dioxide adds to the greenhouse effect that is steadily warming the Earth's climates.

A forest gives out large amounts of water. Much of the water used by Australians comes from forests. It takes 1000 L of water to grow pasture to feed a cow to produce 1 L of milk. Logging a forest can reduce its water output by half.

In the scene at right, a section of eucalypt forest has been cleared for pasture. This will contribute to global warming, as the trees no longer soak up carbon dioxide and give out water vapour and oxygen. The forest that remains in the background is cloaked in mist, while the Sun has burnt the pasture grasses dry. (The cows grazing the pasture will burp methane gas that will add to the gases that cause global warming.)

The endangered Leadbeater's Possum and Sooty Owl depend on old-growth forests for survival.

THE IMPORTANCE OF OLD-GROWTH FORESTS

An old-growth forest has not been logged, or cleared and regrown. It contains trees at many stages of growth, including big, old trees, small young saplings and dead trees. This forest has a large range of habitats, including tree hollows vital to mammals and birds. It may take 1500–2500 years for logged old-growth eucalypt forest to regrow enough to provide a full range of habitats.

FACTS 'N' FIGURES FILE

- **Only 20% of Australia's** remaining forests have not been heavily logged.
- **If clearing of private land** continues at the same rate as today, almost all unreserved forests will have been lost in 170–250 years.
- **Drugs from rainforest plants** are used to treat cancer, AIDS[G], motion sickness, stomach disorders and other conditions.
- **Over half of Australia's biodiversity** is in its forests and woodland, including 53% of its vertebrate animals and 76% of its plants.
- **Australia's forests and woodland** contain more plant species than all of Europe.
- **Over half the wood** logged from Australia's native forests each year is made into woodchips and exported.
- **Over 6 000 000 t** of native forest woodchips go to Japan each year.
- **In March 2001, Western Australia's** Labor Government stopped logging of old-growth forests. Timber workers are being retrained for other jobs.

Humans and forests

PLANTATIONS

Because there has been so much logging of native forests, they now cost more to log and they produce less good timber. Plantations of native and introduced trees produce over half of Australia's domestic wood supply, including two-thirds of the sawn timber. Australia has almost 1 000 000 ha of plantations for timber production. This area is expected to increase to 3 000 000 ha by 2020.

This plantation consists of introduced poplars. They will be harvested and made into matches.

Trimming branches from a Hoop Pine so it can be snigged out of the plantation by a tractor.

PEOPLE ENJOY FORESTS

Forests are important to tourism. In 1998, Australia's tourist industry made 10 times the dollars made by all wood-based forest industries. The Koala depends on forests for survival. It brings over $1 000 000 000 a year in nature-based tourism.

The CALM walkway in West Australia's Valley of the Giants gives a close-up view of the forest canopy.

WORLD HERITAGE FORESTS

At present (September 2001) Australia has 14 places on the World Heritage List. Six of them are forests or contain forested elements. They are:

- The Tasmanian Wilderness
- The Central Eastern Rainforest Reserves of Australia
- Kakadu National Park
- Fraser Island
- The Wet Tropics of Queensland
- The Greater Blue Mountains Area

The Wollemi Pine, an ancient conifer, was discovered in the Blue Mountains in the 1990s. The "dinosaur pine" helped place the Blue Mountains on the World Heritage List.

Glossary

Adhesive Something that makes things stick together.

AIDS Acquired Immuno-Deficiency Syndrome. A fatal disease caused by a virus.

Aromatic Having a sweet or spicy scent.

Botanists Scientists who study plants.

Buttresses Props or supports.

Canopy Leafy branches of forest trees.

Carnivores (*adj.* **carnivorous**) Organisms that eat animal flesh.

Community (*pl.* **communities**) A group of organisms living together and interacting with each other.

Compound Of leaves, divided into two or more parts.

Crustacean Joint-legged marine animals having an outside skeleton and body divided into working sections.

Detoxifies Gets rid of a poison or its effects.

Embryo Organism in the early stages of development.

Endangered In danger of disappearing forever.

Environment The biological conditions in which an organism lives.

Epiphytes Plants that grow on other plants but do not get food or water from them.

Estuaries Places where rivers flow to the sea and are affected by tides.

Evolve To develop gradually.

Extinct No living examples still exist.

Feral Gone wild.

Fertile Capable of reproducing its kind.

Fire regime System of burning country to control its plant growth.

Flora Plant life.

Fossil Remains or impressions of an animal or plant from a past age.

Glands Organs or tissues that secrete substances.

Herb A non-woody plant, such as a grass.

Invertebrates Animals that do not have a nerve chord or a backbone.

Leased Rented for a certain time.

Mammals Animals with backbones whose bodies create their own heat and that have hair on their skins and feed their babies on milk.

Marsupials Mammals whose young are born at an early stage of development and may complete their development in a pouch.

Mature Completely developed.

Membranes Thin sheets of animal or plant tissue.

Monsoon The wind that brings the rainy season (the Wet) to northern Australia.

Nectar The sweet liquid produced by flowers.

Nitrogen An element present in the atmosphere and in animal and plant tissue, chiefly as proteins.

Nodule A small knob or lump.

Nutrient Something used as food.

Nymphs The young stages of insect species that have steady growth rather than dramatic changes of form.

Organic Of living things.

Parasites Organisms that live on or in the bodies of other species (hosts), taking nourishment and causing harm.

Phyllodes Flattened leaf-stems that function as leaves.

Pollinate To carry pollen to the female part of a plant.

Pollution Things that make the environment unclean, foul or unhealthy.

Placental mammals Mammals whose young develop inside the mother's body, nourished by an organ called a placenta.

Predator An organism that preys upon (kills and eats) other organisms.

Recycle To use again.

Regeneration Growing again.

Reproduce Of organisms, to make another of the same kind.

Reptiles Animals with backbones whose bodies take the temperature of their surroundings and whose skins are covered with scales.

Secrete Produce a liquid.

Simple (Of leaves) not divided into parts.

Spores Small reproductive bodies produced by ferns and mosses.

Tannins Vegetable products used to tan leather and used as dyes.

Termites Soft-bodied insects that live in large groups and eat wood and other vegetation.

Torpid A state entered by an animal in which body activity slows down in order to survive a cold period.

Toxins Poisons produced by living organisms.

Transpire (Of plants) to give off water vapour through a surface.

Unique Of which there is only one.

Vapour A substance in the state of a gas, sometimes able to be seen as fog, steam, smoke.

Vessels Tubes that convey some body fluid or sap.

Vulnerable Likely to become endangered if conditions do not change.

A Common Brushtail Possum in a cycad.

Index to subjects illustrated

Some information sources

Websites: Below are a few websites that are full of interesting information. There are many more. Just type into your search engine some key words, such as "australia eucalypt forests", and prepare for information overload.

Australian Museum www.austmus.gov.au
Environment Australia www.ea.gov.au
National Forest Inventory www.brs.gov.au/nfi/forestinfo
National Association of Forest Industries www.nafi.com.au
State Forests of NSW www.forest.nsw.gov.au
The Wilderness Society www.wilderness.org.au

Books: Search for these books in your local or school library. Look in the Contents and Index, check the subject in which you are interested, then find the pages on which they appear. There are many more books on Australian forests. Check the newspapers for current-events information.

Adam, Paul, 1994. *Australian Rainforests*. Oxford University Press, Melbourne.

Beattie, Andrew J. (Ed.), 1995. *Australia's Biodiversity: Living Wealth*. Reed New Holland, Sydney.

Breeden, Stanley, 1999. *Australian World Heritage Tropical Rainforest*. Steve Parish Publishing, Brisbane.

Cogger, Harold G., 1999. *Reptiles and Amphibians of Australia*. Reed New Holland, Sydney.

Gray, Alan T. & Hall, Anne (Eds). *Forest-friendly Building Timbers*. Earth Garden Books, Trentham, V.

Greig, Denise, 1999. *Field Guide to Australian Wildflowers*. Reed New Holland, Sydney.

Holiday, Ivan, 1989. *Field Guide to Australian Trees*. Reed New Holland, Sydney.

Jones, David L., 1986. *Rainforest Plants of Australia*. Reed New Holland, Sydney.

Morcombe, Michael, 2000. *Field Guide to Australian Birds*. Steve Parish Publishing, Brisbane.

Nevill, Simon & McQuoid, Nathan, *Guide to the Wildflowers of South Western Australia*, Simon Nevill Publications, South Fremantle, WA.

Recher, Lunney & Dunn (Eds),1992. *A Natural Legacy: Ecology in Australia*. Maxwell Macmillan Publishing Australia, Sydney.

Reid, Ian G., 1994. *The Bush: A Guide to the Vegetated Landscapes of Australia*. UNSW Press, Sydney.

Slater, Peter, 2000. *The Slater Field Guide to Australian Birds*. Reed New Holland, Sydney.

Smith David, 1994. *Saving a Continent*. UNSW Press, Sydney.

Strahan, Ronald (Ed.), 1995. *The Mammals of Australia*. Reed New Holland, Sydney.

Gum blossoms provide food for many forest residents.

Eastern Grey Kangaroo.

First published in Australia by Steve Parish Publishing Pty Ltd
PO Box 1058, Archerfield, Queensland 4108 Australia

www.steveparish.com.au

© Copyright Steve Parish Publishing Pty Ltd

ISBN 1 74021 088 3

Photography: Steve Parish

Additional photography: Stanley Breeden – Gympie Stinger, lawyer vine p. 13, Southern Cassowary, earthworm p. 16, Lace-lid Frog p. 19, Brown Tree snake p. 21, Stinkhorn fungus p. 39, Blossom Bat p. 40; Pat Slater – Freshwater Crocodile p. 19; Peter Slater – Bell Minor p. 4, Buff-breasted Paradise Kingfisher p. 16, Little Shrike-thrush, Wompoo Pigeon p. 18, Rufous Tree-creeper p. 35, Mistletoebird p. 38, stick insect, Pacific Baza, Little Eagle p. 42; Raoul Slater – Cypress Pine p. 36; Belinda Wright – Musky Rat-kangaroo p. 40.

Text by Pat Slater, SPP
Designed by Leanne Staff, SPP
Edited by Kate Lovett, SPP
Film by Inprint Pty Ltd, Brisbane, Queensland

All rights reserved. No part of this publication may be reproduced, stored in a retrieval system, or transmitted in any form or by any means, electronic, mechanical, photocopying, recording or otherwise, without the prior permission in writing of the Publisher.

Designed and produced in Australia at the Steve Parish Publishing Studios